NO HANG-UPS III

JOHN CARFI & CLIFF CARLE

Illustrated by Greg Tenorio

CCC Publications • Los Angeles

CCC Publications
21630 Lassen Street
Chatsworth, CA 91311

Manufactured in the United States of America

Cover design © 1988 CCC Publications

ISBN: 0-918259-12-6

First printing - April 1988
Second printing - July 1988
Third printing - March 1989
Fourth printing - August 1990
Fifth printing - October 1990
Sixth printing - December 1991
Seventh printing - October 1992

If your local U.S. bookstore is out of stock, copies of this book may be obtained by mailing check or money order for $3.95 per book (plus $2.50 to cover sales tax, postage and handling) to: CCC Publications; 21630 Lassen Street, Chatsworth, CA 91311.

CONTENTS

INTRODUCTION

For Volume III we have used the name PAT (you will, of course, substitute your own name when putting a message on your machine) in all of the messages in this book where a name is required. Why PAT? Well, it can be either male or female. (And if your name *is* PAT, hey, you're all set!)

In most instances, the genders "he" and "she" are interchangeable. Also, if applicable to your status, nouns such as "wife" can be switched to "girlfriend"—or "boyfriend" to "husband"; etc. (Hermaphrodites: ignore this paragraph!)

You might notice that many messages do not contain the generic answering machine phrase, "leave your name and number at the sound of the tone . . . " And with good reason:

1) Answering machines have been around a long time now and only aborigines and people who have been in a 10 year coma don't know what the procedure is.

2) Many people don't like to be told what to do. This world is becoming overcrowded with "sensitive" types who will hang up on your machine at the slightest provocation. You see, the purpose of this book (and our previous books) is to substantially decrease *hang ups*. Since our first book, we have discovered that a funny or clever message alone usually guarantees a response. But, if you insist, the generic phrase can be easily worked into most of the messages at either the beginning or the end of your spiel. Whatever. Afterall, these are now *your* messages, to use or abuse to your heart's delight.

—John Carfi & Cliff Carle

i

Well, here we go again . . .

FOOL YOUR CALLERS

If you're of a rare breed, a person who really doesn't care about **hang ups**—because you're so self-assured you instinctively know that the really **important** people will leave a message on your machine, no matter what, and the people who hang up aren't worth your time, anyway—then this section is for you.

[NOTE: As there are so few such well-adjusted people, this is a short section]

Hello. If this is the first time you've called an answering machine—relax—and I'll give you a couple of pointers.

First—never hang up on a machine or you could receive a severe electrical shock.

And Second—answering machines generate a tremendous amount of static electricity in your clothing—like dryers—so be sure you are naked when you leave your message.

Okay, here comes the beep—hurry up and take off all your clothes!

BEEP. . .

Hi. Here's what just happened:
you put your receiver next to your face—
dialed my number and my phone rang—
so I guess we could say. . . "your face rings a bell!"

But I can't think of your name—so leave it at the tone.

BEEP. . .

Hi. We're not home. We went out to get some change. My wife and I have to flip a coin to see who will be on top. Boy, I hate putting a new roof on a house!

BEEP. . .

(WHISPERING)
You won't believe this, but I've just been visited by Aliens! Yeah, no kidding! As a matter of fact, they come every week—and clean my house. The leader's name is Maria. The other is Juan.

BEEP. . .

Hi. PAT here. Guess what? I bought you a present. That's right, a brand new answering machine! You're listening to it right now. So what do you say? Why not try it out? It's yours. . .

BEEP. . .

I'm not home. I'm over at my parents again. Last time, my Mom made me *sole* food—then started yelling at me, saying I was a real *loafer*. And my father called me a *heel*. Hey, you don't like my message? So *shoe* me!

BEEP. . .

You know, they say we use not even one-fourth of our brain. So the way I see it, anything you say is only worth a quarter.

BEEP. . .

Congratulations caller! You've been voted "Sucker Of The Year!" At the tone, please leave your name and all your credit card numbers.

BEEP. . .

[MESSAGE FOR FEBRUARY 2]

Hi. I went to the market to buy some sausage—
it's a tradition with us—afterall, it is Ground-*hog*
Day!

BEEP. . .

(ANGRY)
I can't believe this! I wait by the phone all day
for your call—nothing! Then I step outside for a
minute to check my mail and what happens???
(ELATED)
I win a free vacation to Hawaii! Leave a message
and I'll get back to you in a couple-a weeks!

BEEP. . .

(PERPLEXED)
Uh, gee, sorry. I had a real funny joke for you
today, but now I can't seem to find it. I must have
left it in my *other* machine???

BEEP. . .

At the tone, please leave a **complete** message because there are **three** things I can never remember—
One is names.
Two is numbers.
And **Three** is. . . Uh??? Is. . . Uh???

BEEP. . .

Hello. This is today's message. If you want to hear **yesterday's** message, just call back tomorrow.

BEEP. . .

Hello. You have reached The Society Of Re-repeating Things Over And Over Again Redundantly A Lot. Please kindly leave your name who you are and the digital integers of your phone number listing when you hear the sound of the beep tone noise. Thank you ever so very much! Beep!

BEEP. . .

Hi. Sorry if you haven't been able to reach me lately. I just got back from a Borneo Safari where we hunted wild pigs. Leave a message—I'll call you back and *boar* you with the details.

BEEP. . .

Hi. If this is a member of the opposite sex call-ing, I have a question for you?
Are you free tonight. . . . . . .
Or is it going to cost me?

BEEP. . .

(RELIEVED)
Oh, thank goodness you finally called! I've been trying to reach you all day, but I lost your number. So leave it at the tone and I'll call you right back!

BEEP. . .

Hi. This machine is divided into two sections: *First Class* and *Coach*. For Coach, just leave your name and number. For First Class, go get a glass of champagne—then leave your name and number.

BEEP. . .

I used to be home *more*. But now I'm home *less*— because I have a heavy social life— which, by the way, is costing me a fortune. So, at the tone, leave a message and if you can spare it, please send money to the *home less*.

BEEP. . .

Hello. This is PAT. I was just looking at the ERASE button on my machine and it got me thinking—now, I don't want to get too metaphysical or anything—but where the hell do words go when you erase them???

BEEP. . .

(EXCITED)
Hi. This Is PAT. Guess what? This morning I was on the radio! And this afternoon I was on the TV! But my rear-end is getting sore, so tonight I'm going to be on the sofa!

BEEP. . .

Hi. We went out for dinner. Hope it turns out better this time! Yesterday we went to this lousy seafood restaurant. What a night! The employees were rude—the customers were mad—and the clams were steamed!

BEEP. . .

Leave a message. I'm running late for my Ego Club meeting—but not to worry, they wouldn't think of starting without me! Today we're voting in a new president, so I had to rehearse my acceptance speech. Well, gotta go—it was nice talking to me.

BEEP. . .

THE YOLK'S ON YOU

In getting someone to leave a message, there's one thing that is sometimes just as effective as humor. . . . *guilt*.

And probably the best way to achieve guilt is through sympathy. The theory here is, if you get the caller to feel sorry for you, they will usually leave a message. Or, if they hang up, sooner or later they will begin to feel guilty and will have to call you back. You will know it's working when you play back your day's recordings and find the pattern:

HANG UP/MESSAGE. . . HANG UP/MES-SAGE. . . HANG UP/MESSAGE. . . etc.

And you will know you're really getting good when you find that certain callers have felt *so* guilty, they called back twice, using different "cheery" voices!

Hi. We're not answering the phone—we're all feeling kind of *ugly* today. See, we had a family portrait done in oils—a friend saw it and said, "Oh, is that a Picasso?"

BEEP. . .

(EXCITED)
Hi. I'm on my way over to City Hall. I'm so proud! I think I got a new job as an advisor to the Mayor—yeah, his exact words were, "If I want your advice, I'll *ask* for it!"

BEEP. . .

(SMUG TONE)
Hi. This is PAT with a tip for you. You know how, when you're out of town and you call your home machine to get your messages by remote, it's very expensive? Well, do what I do—reverse the charges!

BEEP. . .

Hi. Leave a message. I'm out on a date. I'm dating this writer. You know, it's tough dating a professional writer. Like, last week he sent me this sweet, romantic love poem—then he billed me for it.

BEEP. . .

Hi. I'm at the Post Office. I think I put too many stamps on a package last week and I'm worried that it may have gone too far!

BEEP. . .

Hi. I'm down at the gym working out. Yesterday, my exercise instructor said I was "ship-shape". I thought that was good news till I found out he meant I hadda lose 20 pounds on my port side—another 15 astern!

BEEP. . .

Hi. I went to the doctor. I think I have kleptomania and I gotta find out what to *take* for it.

BEEP. . .

Hi. Please leave a message. I'm out looking for antiques. I'm really into it! I think antiques are great—that is, if they're not too old!

BEEP. . .

(UPSET)
Hi. I went to see a lawyer. A couple of days ago I bought a see-through negligee. The next day, my husband bought a sleeping mask.

BEEP. . .

[XMAS MESSAGE]

Hey, the holidays are upon us and I'm so excited about getting into all those Xmas traditions! You know, decorating the eggs, writing Valentine cards, lighting the firecrackers, dressing up in those scary costumes! But I have to be careful—'cuz if I see my shadow, I gotta go back into my basement for six more weeks!

BEEP. . .

(DEPRESSED)
Hi. Leave a message. I'm out trying to sell my brand-new sports car. I have no choice. It's the only way I can pay for the insurance on it.

BEEP. . .

(UPSET)
I knew I shouldn't have laid my answering machine on top of my tape-deck! I come home nine hours later and what do I find on the floor? All these little mini-cassettes!
Anyway, can ya help me out? I'm trying to find a home for them!

BEEP. . .

Hi. I'll be back in a while. I had to go in for my yearly eye exam. My eyesight is so bad, when I have to read the eye chart I start playing **Wheel Of Fortune**. . . "Is there a *G*? — I'd like to buy a vowel!"

BEEP. . .

(WEAK VOICE)
Hi. I'm home, but I can't pick up the phone. I just got back from the shopping center. A word to the wise: never walk into a health food store while eating a hotdog! I'm in bed with severe 'granola wounds' all over my body!

BEEP. . .

Hi. I won't be picking up this morning—I have sort of a weird hang-over. Last night I went to one of those all-you-can-eat salad bars—guess I had one too many "mixed vegetables". Now I have to go to my "AA" meeting—you know, Asparagus Anonymous!

BEEP. . .

Leave a message. I'm over at my new girlfriend's. We just met at a bar. You know, they say you never meet nice girls in bars. But she's warm, sensitive and very understanding, too. We made love for hours and you won't believe this: she only charged me fifty bucks!

BEEP. . .

I won't be answering the phone for a while because I'm having my teeth cleaned—that's not so bad—except I won't get them back till Friday.

BEEP. . .

(MOROSE)
I can't pick up the phone right now. I'm too depressed. Financially, I'm completely tapped out and my career's going nowhere. You know that show starring Robin Leach? They should have a show for people like me—call it "Lifestyles of the Poor and Insignificant".

BEEP. . .

Please leave a message. We're off on another trip. You know, I think I'm more afraid of flying than most people. I'll be sitting there with my seatbelt on—sweating, shaking, biting my nails. Finally, my wife will say, "Honey, get out of the car! You'll miss your plane!"

BEEP. . .

Hi. I went out for another drive in the country. I'm taking a different route this time! Yesterday, while I was driving, I came to a sign that said "*Draw*bridge". Ticked me off! It must-a took an hour before I could find a pencil!

BEEP. . .

(MAUDLIN)
Hi. Please leave a cheery Message. Nothing's going my way lately. I'm so unlucky—first, I get a parking ticket at the Drive-In. And then, when I get home, my swimming pool's on fire!

BEEP. . .

I'm not in. I left for my doctor's appointment. If you know a good doctor, let me know. Mine is really mean. Last time he said, "Tell me if this hurts"—then showed me his bill.

BEEP. . .

Please leave a message. I'm over at my new property. You know, I think I'm finally getting the hang of this real estate investment thing. I bought this house—it's not exactly 'water front', but the realtor guaranteed me that, next tidal wave, it will be!

BEEP. . .

Hi. I'm taking off. I just realized I haven't eaten all day.
(TO SELF)
On second thought, I think I'll skip dinner—I hate to eat on an empty stomach.

BEEP. . .

Sorry, I'm not in. I went to the Drug Store. For about a month I've been using this Right Gaard deoderant. It works great, but I gotta find out what to do about my *left* arm!

BEEP. . .

I can't come to the phone—I'm sick in bed. They had to pump my stomach. This morning I ate "All Natural Trail Mix"—in my stomach they found granola, raisins, empty beer cans, some cigarette butts and part of a pup tent. Funny—I always thought a little *fiber* was good for ya???

BEEP. . .

Hi. Maybe you can help me. I can't figure out what kind of ticket I need. I'm going to take a trip around the world—do I get a round trip ticket? Or one-way???

BEEP. . .

Hi. Leave a message. I had to go back to the "Stop Smoking Clinic"—it's one of those places that use shock therapy. They cured my smoking, but now I'm addicted to sticking my finger in light sockets!

BEEP. . .

I'm out. I had to go over to my doctor's to get my test results. I wanted him to give it to me over the phone, but he doesn't make house calls.

BEEP. . .

Hello. This is PAT. Please leave your name and. . .
(PANIC)
Oh my god! I'm looking out the window here and someone is actually stealing my car! I can't believe it!
(RELAXED)
Never mind—it's okay—I got the license number.

BEEP. . .

Hi. I'll be back later. You know, I have to either lose weight or find a new dry cleaner. I took my dress in this morning and they said they don't do tents!

BEEP. . .

(LOW KEY)
Hi. I won't be picking up the phone the rest of the day. Ever since I accidentally swallowed some developing solution I've been feeling kind of negative—and I can't seem to get rid of this film on my teeth!

BEEP. . .

Hi. I'm out—got a great new hobby! I'm into "antiquing"! Yesterday, I bought this real old Coke container for 500 bucks. I know it's genuine because right on the can it said "Coke Classic"!

BEEP. . .

Sorry I didn't get back to you last night, but I got in too late. I went and saw this really horrible movie. It was so bad, in order to get my monies worth, I had to watch it *three* times!

BEEP. . .

I'm not home to answer the phone. You probably won't believe this, but I'm in jail. I got arrested for indecent exposure while I was in the Laundro-mat. It wasn't my fault—absolutely everything I owned was dirty.

BEEP. . .

(WEAK VOICE)
Hi. Leave a message. I'm in bed with an upset stomach. I just ate two pizzas, three cheeseburgers, two malts, large fries and two pieces of apple pie a-la-mode with a cup of coffee. Darn! I knew I shouldn't have drank that coffee!

BEEP. . .

Hi. I'm out looking for a new broker. My last broker put half my money in toilet paper and half in revolving doors—I was wiped out before I could turn around!

BEEP. . .

Hi. I went over to my travel agent to try and get my money back. I felt great about this economy cruise I signed on for until I got their list of "essentials" to pack for the trip: lifejacket—bailing bucket—and a pair of oars.

BEEP. . .

(EXCITED)
Hi. I'm not going to be home for a while. I finally got my big debut on TV! You can see me on the six-o-clock news! Look for the guy in front of the "7-11" with the pantyhose over his head!

BEEP. . .

(WHISPER)
I can't answer the phone—I'm in shock. Last night when I was out drinking, I met this girl who looks like a million bucks—this morning when I woke up, she looks like a bounced check!

BEEP. . .

Hi. Leave a message and I'll return your call— if I still have a phone. I got *another* outrageous phone bill. I think this time I'll just send them the deed to my house and be done with it.

BEEP. . .

(EXCITED)
Hi. I can't come to the phone. I'm totally engrossed in this new novel I just picked up. It's called. . . let's see. . . where's the title??? Well, I can't find it, but what's amazing is, all the words are in alphabetical order!

BEEP. . .

In this section, covering a wide range of subjects and situations, there will be a message (or several messages) for just about everybody.

When we finished this section, we thought and we thought and we thought, but for the life of us, we just couldn't come up with a heading. Then, suddenly, we both arrived at the exact same conclusion:
That's why they invented the word

MISCELLANEOUS

Hi. I'm out. I went to visit my neighbor in the hospital who's recuperating from brain surgery. The guy is dogmatic, stubborn and very opinionated. To tell you the truth, this is the first time in his life he's had an open mind.

BEEP. . .

Hi. I'm out with an old girlfriend. She's really great looking now, but when we were in Grade School she was ugly. I mean really ugly! One time, for a couple weeks, she was missing and they put her picture on a can of Alpo.

BEEP. . .

Sorry, I can't pick up the phone. I'm disciplining **All My Children**. I don't want them to take after me and watch too much TV. I want them to grow up living **Lifestyles Of The Rich And Famous**. I mean, **Gimme A Break**, they only have **One Life To Live**—and remember, **Father Knows Best**!

BEEP. . .

Hi. This is PAT. You know, Hula-Hoops are out, Bell Bottoms are out and I'm out. So leave your number and I'll call you back.

BEEP. . .

I'm not home. I'm visiting this girl I met at a singles' bar last night. She has a great attitude! She says everyone should practice "safe sex". Anyway, we'll be practicing all day.

BEEP. . .

Please leave a message. I'm out looking for my wife. She's been gone for three weeks. I don't know if she's lost or she just went shopping!

BEEP. . .

(FAST)
Hi. I'm running down to the store. I just saw an advertisement on TV for the all new "Gorbachev Lee-press-on Birthmarks"! Gotta hurry before they run out!

BEEP. . .

Hi. I can't answer the phone. I have a crisis here. My parents are staying here for a few days and last night they came home after 2:00 a.m. I don't know what to do—ground them? Take away the TV? Maybe what they need is a good old-fashioned spanking!

BEEP. . .

Hi. I can't answer the phone right now 'cuz I'm too hungry. You know what would taste real good right now? Some tri-sodium casenate with di-potassium phosphate, silico-aluminate, thiamine mononitrite and sodium benzoate. . . I mean a *cupcake*.

BEEP. . .

[MESSAGE FOR A RAINY DAY]

Boy I hate this! I have to go out and it's raining cats and dogs. It's not the downpour I mind— I'm allergic to animals!

BEEP. . .

Hi. I thought you might be interested in a little telephone trivia:
Alexander Graham Bell invented the first telephone and successfully used it every day for a month. As a matter of fact, he was so excited about his new invention, he forgot to pay his phone bill and it was disconnected.

BEEP. . .

Hi. This is PAT. I'm out on a date. I'm seeing this new guy. He's so *macho* he smokes in bed— face down.

BEEP. . .

Hi. I'm visiting my uncle. The guy's a little overweight. As a matter of fact, the other day I was looking through the family album and in every picture of him, there's a refrigerator in the background.

BEEP. . .

Hi. I'm back in town. Sorry if you missed me last week. I went to my class reunion. You know, some people never change. There was this guy in high school who was really sleazy—he's still the same—his date for the reunion was an inflatable doll!

BEEP. . .

I'm not home. I had to go visit my cousin. The guy's a mess! He just found out he has a split personality. I'm telling you, he's really beside himself!

BEEP. . .

[HALLOWEEN MESSAGE]

Hi. I won't be picking up. I'm a little hung over from a costume party last night. Great party! This one girl was completely naked except for black gloves and black shoes—she came as the "five of spades".

BEEP. . .

Hi. I'm not home. I went to check out this really weird market that just opened. It features stuff like **head** cheese, chicken **hearts**, **leg** of lamb, **arm** roasts, pig's **feet**, beef **brains**—the place's called "Frankenstein's".

BEEP. . .

Hi. I can't pick up. My nephew was screaming and yelling and now he's locked himself in the bathroom. He doesn't want to go to camp because it starts on **Friday the 13th**. I think I'm going to have to have a word with the new camp counselor, **Jason**. . .

BEEP. . .

Hey, I'm back from my vacation. It was great! One day I had this great dinner in an Old West Saloon where they served bison steaks—then after the meal, they give you the buffalo bill.

BEEP. . .

Hi. I'm not in. I'm down at the corner phonebooth making prank calls. Here's one you can try: look up the name 'Booger' in the phonebook. If the guy's name is 'Bob', ask for *Mike*. The guy will say, "There's no one here named 'Mike'." You say, "Sorry, I must have picked the wrong *Booger*!"

BEEP. . .

I'm out. I went to a birthday party for this friend—I hate to say it, but the guy really has bad breath. Anyway, a bunch of us guys, we're all chipping in to buy him a keg—of mouthwash.

BEEP. . .

Hi. I have to go to a wedding today. By the way, I was just thinking—you know how *we* throw rice at weddings? Do you think Far Eastern people throw hotdogs?

BEEP. . .

Hi. I went to visit my aunt, who's ***really fat***! It's kind of interesting—people follow her everywhere. She's not popular, it's due to her 'gravitational pull'.

BEEP. . .

Hi. This is PAT. Because I'm against hunting I went to a protest meeting of the "Deer Retaliation League". It seems to be working. Today we saw a buck with a hunter strapped to the front of his antlers.

BEEP. . .

(UPSET)
Please leave a message. I gotta bail my brother-in-law outta jail—reckless driving! He claimed he hadn't had a drop—was in full control. But when they gave him a urine test, he missed the cup.

BEEP. . .

(CHUCKLING TO SELF)
Hi. I'm going out again. Boy, did I have a good laugh today! My Ex had me over because she spent ten thousand bucks on an abstract painting and she wanted me to see it—turns out it was a mirror!

BEEP. . .

Hello. You should know: I wired my answering machine to my VCR and you are being video-taped right now, so I'll know who you are. Don't believe me? Okay, I'll prove it: If you hang up, I won't call *you* back!

BEEP. . .

Please leave a message—I went over to the hospital. I have a friend who went in for breast enlargement—and boy is his wife pissed at him!

BEEP. . .

[ACTOR'S MESSAGE]

Hi. I'm at my acting class. Last week my acting teacher said, "If you sleep with me, I'll get you a starring role in a movie." I said, "Why don't you get me the part and just *act* like I'm sleeping with you!"

BEEP. . .

Hi. You'll have to leave a message. I'm taking my Cam-Corder and going out to a bar. It was my wife's idea—she told me I should *watch* my drinking!

BEEP. . .

Winners Of
CCC PUBLICATIONS
NATIONAL
"BEST MESSAGE"
CONTEST
(July 1, 1986 - March 31, 1987)
[NOTE: minor editing on some messages]

FIRST PRIZE ($1000) WINNING MESSAGE:

Peter J. Nemarich
New Haven, Connecticut

"Your eyelids are getting heavy. . . you are suddenly very, very tired. You're entering a D-E-E-E-P sleep. . . thinking of nothing but the sound of my voice. Tonight at midnight, you will get up, and you will write me a check in the amount of your account balance. You will mail the check to me in the morning, on your way to work. When you hear the beep, you will awake and remember nothing."

BEEP. . .

SECOND PRIZE ($100) WINNING MESSAGE:

Shelly K. Scott
South Mission Beach, California

Although I like to do it up and down, my boyfriend prefers to do it back and forth. But either way we both achieve maximum results and are left with a wonderfully ALIVE feeling. So as soon as we've finished brushing our teeth, we'll return your call!

BEEP. . .

THIRD PRIZE ($50) WINNING MESSAGE:

Geralyn L. Ashum
Orland Park, Illinois

Hello, I was expecting your call! That is why I left my answering machine on. That is also why I'm not home!!

BEEP. . .

FOURTH PRIZE ($5)
WINNING MESSAGES

(IN ALPHABETICAL ORDER)

John M. Arnone
Peoria, Illinois

Hello. I **am** home right now, but leave your message after the beep anyway, because this machine is my only form of entertainment.

BEEP. . .

Michael Samuel Aurelius
Munster, Indiana

"Dis ist Wolfgang Mozart. You interrupted my decomposing. **NAME** ist nicht zu Hause. He vent out to look at some music. . . vell, he told me he vas going to check out some bars."

BEEP. . .

David M. Beach
Cincinnati, Ohio

(SOUTHERN PREACHER STYLE)
Hello friend and welcome. Please listen to my words. Please leave me some form or sign of your calling. Because the "Big Operator" called me and said I need at least three million calls immediately or my line will be permanently disconnected. So, please be **oral** when you hear the tone. Bless you.

BEEP. . .

Terri Black
Beaverton, Oregon

Studies indicate that people who hang up on answering machines are insecure, paranoid, and psychotic. People who talk to answering machines are secure, intelligent, and successful. Please categorize yourself after the tone.

BEEP. . .

Robert J. Blake
New York, New York

I'm sorry, there's no one here who can talk to you right now. Marcel Marceau is here, but he never says anything. Harpo Marx is here, but he's not saying a word. Joan Rivers and Johnny Carson are both here, but they're not speaking. So, at the tone, YOU speak!

BEEP. . .

Kay Bienvenu
East Greenwich, Rhode Island

(MONOTONE)
Hello. There is nobody home. At the sound of the tone, please leave your name and number. . . Hey! You want jokes and silly patter? Go to Las Vegas. I am only a machine.

BEEP. . .

Sarah E. Borgman
Waharusa, Indiana

(SING)
"From the halls of Montezuma,
To the shores of Tripoli. . ."
(SPEAKING VOICE)
Like the Marines, I, too, am looking for a few
good men. If you qualify, leave your name and
number and I'll get back to you.

BEEP. . .

Dorothy Papalia Castignetti
Glenview, Illinois

(HIGH PITCHED VOICE)
Hey there! This is your old friend, Mickey. The
Mouse Club is closed right now. When Jimmy
gives a whistle, please leave your name, number
and a message and I will call you back. Why?
Because I *like* you. . .

BEEP. . .

45

Craig Dickens
Evanston, Illinois

BASED ON "STOPPING BY THE WOODS ON
A SNOWY EVENING"
BY ROBERT FROST:

Whose phone this is, I think I know
his house is in the village though
He will not know I called him here
if I choose not to tell him so—
My little cat must think it queer
to speak when there is no one near
A Snoopy phone clutched at my side
it's receiver pressed against my ear—
I missed him but I will not weep
for he has promises to keep
And calls to return before his sleep
I'll leave my message at the beep—

BEEP. . .

Ann L. Dwyer
Northbrook, Illinois

VOICE #1: (TO SELF) So, how's this thing work, anyway???

VOICE #2: Hey, what're you doing?

VOICE #1: Recording a message.

VOICE #2: Just turn the knob and press the button.

VOICE #1: What button???

VOICE #2: *That* button!

VOICE #1: Yeah, but how do we know if it's recor. . .

BEEP. . .

Ron Elkins
Evanston, Illinois

In the interest of telephone safety, please look both ways before leaving your message.

BEEP. . .

Peggy Englebert
Lansdale, Pennsylvania

Hi. This is **NAMES'S** refrigerator. Her answering machine isn't working right now and I'm trying to fill in. So . . . when I open my door . . . and the light goes on . . . leave a message. Uh, is that how it works???

BEEP. . .

Barbara Rainbow Fletcher
Kirkland, Washington

This is NAME speaking to you by machine. I'm on vacation so I can't come to the phone. Let me give you a word of advice before you leave for the holidays. *Never* go anywhere in a car where the kids outnumber the windows! If I survive I'll call you as soon as I return.

BEEP. . .

Philip Gardocki
KIng Of Prussia, Pennsylvania

Greetings, Earthling. I am a creature from outer space. I have transformed myself into a telephone answering machine. Right now I am having sex with your ear. I can see you are enjoying it, because you are smiling. . .

BEEP. . .

Georgia L. Geis
Mesa, Arizona

(MECHANICAL VOICE)
Hello. You have reached a species of higher intelligence—otherwise known as an answering machine. We are here to annoy you and make you hang up. But if our mission has failed, leave your message at the beep.

BEEP. . .

Karen D. Gintovt
Falls Church, Virginia

Tomorrow, and tomorrow, and tomorrow,
Creeps in its petty pace from day to day;
At the last syllable of this recorded message
Will come a beep, and then you can say,
"I have a message for *NAME*."
Out, out with it then, and pray be not long,
For we do not suffer fools gladly,
Especially those who dial the number wrong.
And when these messages,
Told by idiots, full of sound and fury,
Are heard by us,
We'll call you back in a hurry!

BEEP. . .

Debby Hawkins
Los Angeles, California

Hi. This is **NAME**. Congratulations! I gave you
the right number. Now you have 60 seconds to
convince me to call you back. Good luck.

BEEP. . .

Michelle Heiden
Springfield, Illinois

Hi. This is **NAME**, the sultry blonde with loose
morals and a body that could cause wide-
spread civil unrest, with thighs that could be the
basis for a worldwide religion. I can't come to
the phone right now as I'm either at the Mall
spending all of Daddy's money or I'm admiring
my fabulous bod in the mirror. Please leave your
message at the "moan" and I may get back to
you. Oh, by the way, how did you get this
number?

BEEP. . .

51

Ellen Hildreth
Akron, Ohio

This is **NAME**. I'm having sex right now, so leave your name and number, and I'll get back to you a **lot** sooner than I **used** to.

BEEP. . .

Sharon J. Horner
Sioux City, Iowa

(SOFT, SENSUOUS VOICE)
Ahhhhhh—you called. I go out of control whenever I hear your voice—lusting for you and wanting your body—closer, closer, closer! I can't be satisfied. At the tone, tell me your desires!
(NORMAL VOICE)
My apologies—this machine once belonged to a Dial-A-Porn service.

BEEP. . .

Len Jacobs
Houston, Texas

(BELA LUGOSI'S VOICE)
Good evening. This is Count Dracula. Personally, recorded messages drive me bats, but I just stepped out for a bite. If you don't mind sticking your neck out, just leave your name and number and you can be bloody sure I'll get back to you. . . before sunrise.

BEEP. . .

Scott Jamieson
Portland, Oregon

("STUFFY" VOICE)
Hello, and welcome to dial-a-concerto. Today, we'll be hearing a selection from **NAMES'S** Opus One, Number One, for unaccompanied answering machine; the so-called "One-Note Concerto." As always, your comments are welcome immediately following the performance:

BEEP. . .

Frank Johnson
Cincinnati, Ohio

[MUSIC: "AS TIME GOES BY"]

(BOGART VOICE)
Of all the phone numbers in all the towns in all the world, you call up mine. Well, I'm not in right now. I'm out arranging for the Letters Of Transit. So I guess you'll have to do the talking for both of us. Here's listening to you, kid.

BEEP. . .

Maureen S. Kelly
St. Cloud, Minnesota

Hello. This is WONDERWOMAN!
I WONDER who you are?
I WONDER why you called?
I WONDER what time it is?
And I WONDER if you'll respond when you hear the beep?

BEEP. . .

Mabel I. Ketel
Portland, Oregon

Sorry you didn't reach us. Two of us take turns answering the phone. The one leaving has already left and the one coming isn't here yet.

BEEP. . .

Janie Knowles
Raleigh, North Carolina

NAME'S Laxative Testing Laboratory—**NAME** speaking. Right now I'm busy working my butt off. If you'll please leave your name, number and message, I'll return your call as soon as I'm finished—if I'm not too wiped out.

BEEP. . .

Fred Lenhoff
Oak Park, Illinois

Hello, it's me. If you're who I think you are, leave a message and I'll call you back. If you aren't who I think you are, leave a message anyway. If you aren't who you think **you** are, don't leave a message—go get professional help. If I'm not who you thought I was, leave a message anyway and maybe we could do lunch. **Ciao**.

BEEP. . .

Melvyn J. Loftus
Alexandria, Virginia

I am putting in a cameo appearance at work. Leave a message at the tone.

BEEP. . .

David M. Malkin
Rowland Heights, California

Due to technical difficulties my feet are unable to get my mouth to the phone. Please leave a message

BEEP. . .

Richard D. Manning
Los Angeles, California

[MUSIC: FIFE & DRUM]

(NARRATOR VOICE)

This is **NAME**. Two hundred years ago today, Benjamin Franklin invented the world's first telephone answering machine. "With this device," Franklin said, "you'll even be able to leave a message when you hear the tone." However, Franklin's machine never really caught on—because at that time, the telephone hadn't yet been invented. I'm **NAME**, and that's the way it was.

BEEP. . .

Fred Mathews
Woodland Park, Colorado

[TAP DANCING MUSIC IN THE BACKGROUND]

Hello! I didn't make it to the phone. So please leave your name, number and message and I'll call you back. But be careful what you say. I think someone's tapping the line.

BEEP. . .

Gary Mills
Lexington, North Carolina

Answering machines are a lot like social diseases. . . nobody wants one, but practically everybody has one!

BEEP. . .

John Moran
Phoenix, Arizona

Hi. Thank you for calling. Picture if you will a naked, sensuous and inviting Vannah White, lying on white satin sheets, beckoning you to join her. I pictured it. That's why I'm not on the phone. At the sound of the heartbeat, please leave your name and number and as soon as I catch my breath, I'll return your call.

BEEP. . .

Mike O'Brien
Portland, Oregon

You have reached *NAME'S* car phone. . .
Oh, I see. Just because he drives a little, bitty car, you don't believe it has a phone in it. Well, just to prove it, I will honk the little, bitty horn on his little, bitty car. Listen, now. . .

BEEP. . .

Shannon O'Donnell
Tacoma, Washington

Thank you for calling the "No Deposit—No
Return" Answering Machine. You no deposit
your name, I no return your call.

BEEP. . .

Bonnie A. Piddington
Des Plaines, Illinois

Did you ever have one of those days when
absolutely everything went your way? I mean,
a day when the birds were singing outside your
window when you woke up and all the traffic was
going the opposite direction when you hit the
expressway? A day when your boss actually
gave you credit for your work and the dough-
nuts from the deli were fresh? A day when you
dialed a number, wanting to speak to that special
person and there they were on the other end of
the phone just waiting to talk to you? Did you
ever have one of those days? I hope so—'cuz
this ain't it.

BEEP. . .

Steve Raglin
Omaha, Nebraska

(VERY EXCITED, OBNOXIOUS STEREO
STORE SPOKESPERSON VOICE)
YOUR CITY, this is it! The home of *FULL NAME*!
You're listening to his NEW name-brand electronic answering machine—WITH beeperless
remote. *NAME* has a HUGE inventory of errands
today—however, for a limited time, you can
leave your message—up to *60* SECONDS!
UNBELIEVABLE! Do it NOW!

BEEP. . .

Tom Reilly
Phoenix, Arizona

Welcome to the age of excuses. First it was Nix-
on telling you he was not a crook. Then it was
Reagan telling you he was not aware. Now it is
NAME telling you she cannot come to the phone.

BEEP. . .

David L. Rennie
Buffalo, New York

(AS *LENNY* IN "OF MICE AND MEN")
Hello there. George wanted me to tell you that
the *NAME* family is not at home right now. But
if you will leave a message at the little beep, they
will get back to you. Please leave a message,
because George said he would let me pet the
bunny rabbits for every message.
Thank you.
(AWAY FROM MIKE)
Did I do good, George?

BEEP. . .

Ron Robinson
Pasadena, Texas

Hello. This is *NAME*. We have asked this smart-
aleck machine to receive your message for us.
It insists on giving you a coded clue as to why
we're not doing it ourselves. Three beeps means
we are grocery shopping. Two beeps, we've
gone out to eat. One beep, we are making love
and don't want to be interrupted by the phone.

BEEP. . .

Hal Rosene
Boring, Oregon

Greetings. This is **Rhett**, the family **butler**. I am presently writing **Scarlet** a letter. The Lord and Lady are attending a **fire** sale in **Atlanta**. No matter. Leave your message at the beep. We'll see that it's not **gone with the wind** and we won't **Tara** in getting back to you.

BEEP. . .

Robert L. Sandelman
Brea, California

(MALE VOICE)
Hi. This is Rich Littles, impersonating **JOHN**.
(FEMALE VOICE)
Now I'm impersonating **JANE**.
(MALE VOICE)
I'm doing a pretty good job, aren't I? **JOHN** and **JANE** can't come to the phone right now, but if you'll leave a message, I'll impersonate **your** voice for them too! They may even let **me** return your call! How fun!

BEEP. . .

John W. Simpson
Birmingham, Michigan

Sorry we can't come to the phone. But if you are caller Number 5, you will win two tickets to a Russian dog show. . . or is it the Miss Moscow Pageant??? Well, what's the difference?

BEEP. . .

Michael D. Spencer
San Jose, California

(FAST)
Thanks for calling **NAME'S** MOVIES. Today's program includes "9 to 5" and Fellini's "8½". "9 to 5" plays at 7 and 11, but never at 9. "8½" plays only at 9 and never at 7 or 11. Also showing: "Ocean's 11" and Butterfield 8". "11" shows at 9 and "8" at 7 and 11, Monday through Saturday, but **never on Sunday**. And the matinee any Wednesday between 9 and 5. "Saturday Nite Fever" plays at 6 and 10. "Never On Sunday" plays only on **Golden Pond**. All features repeat at **the same time next year**. Please leave your message to me at the sound of the **shot in the dark**. . .

BEEP. . .

Tom Tozer
Lima, Ohio

(FEMALE VOICE)
Hi. In case this is an obscene phone call, no doubt you are wanting to elicit various emotional responses in order to satisfy your sick and twisted desires. Even though I am out at the moment, I'll do my best: Who is this! . . . That's disgusting! . . . Hey, I don't need to listen to this! . . . You oughta be put away!
(LONG PAUSE)
You can do *what*? Listen, give me your name and number—and don't leave your phone booth. I'll be right over!

BEEP. . .

Paul Varga
Westland, Michigan

Hi. Sorry, I can't come to the phone right now. I have a problem. I just bought some *powdered water* and I'm not sure what to add.

BEEP. . .

Kenneth L. Ward
Wauconda, Illinois

Now stay calm and don't panic. In 20 seconds there will be a short oral exam. If you would like a few quick study tips, I suggest that you start with your name and reason for calling. However, since you have obviously already paid the 25¢ exam fee, feel free to flunk by hanging up.
BEEP. . .

Sheila E. Wenz
Studio City, California

(STEWARDESS-TYPE VOICE — FOR BEST EFFECT, HOLD NOSE)
Hello and welcome aboard **NAME** Airlines. We can't come to the phone right now because the Captain has turned on the **Fasten Seat Belt** sign. Please leave your name and number in its full upright position by grasping the phone receiver and placing it firmly over your mouth and nose. Also, it may be used as a flotation device. Once we reach our final destination, we'll be sure to call you back. Once again, thank you for flying **NAME** Airlines and we hope you think of us for your future phoning needs. Please wait for the beep and assume the crash position.
BEEP. . .

Betty Suddarth Widner
Kokomo, Indiana

(SENSUAL)
Hel-lo-o. It's me, you big, beautiful hunk. Ward took Wally and The Beaver to the lake for the weekend. I'm free for you, Darling. . . free to take off this ugly housedress they make me wear and let the real June Cleaver stand naked before a real man. Come to me as soon as you can. Oh, and Ward said it looked like it might rain later. . . be sure you bring along your rubbers. . . Bye-bye.

BEEP. . .

Ed F. Wildfong
Portland, Oregon

This is the electronic idiot responding to your call. You can give the electronic idiot a message for someone. You can have the electronic idiot ask someone to call you back. Or, you can call again later in the hope that a real live idiot will answer.

BEEP. . .

SPECIAL EFFECTS!!!

[AND SPECIALIZED VOICES]

This section will probably be the most fun—for both you and your callers.

For the following messages, you don't have to be a fifth-degree answering machine professional—daring amateurs will do just fine. Although the degree of difficulty is low, your callers will be amazed just the same.

All of the props necessary for the Special Effects will be found in the average home—no steamship whistles or exotic jungle noises required. However, in some instances, you may need a friend to assist you (finally get some use out of that shiftless roommate of yours!)

Have fun!

PROPS: POWER DRILL
COTTON BALLS

Hi. I'm home, but I won't be able to answer the phone. . .
[STUFF COTTON INTO MOUTH]
Excuse me, I'm putting the cotton in now. . .
(GARBLED)
The kit I ordered just arrived and I can't wait to try it out! It's called "Do-It-Yourself Home Dentistry"!
[REV POWER DRILL]

BEEP. . .

PROP: VACUUM

[VACUUM RUNNING]

(SHOUTING OVER VACUUM)
Hi. I probably won't hear your call 'cuz I got the vacuum going. I thought I'd surprise my wife and do some housework while she's out shopping. She thinks men have no domestic skills! Anyway, leave a message and I'll call ya right back—'cuz I'm just about done here, cleaning out the refrigerator.

BEEP. . .

PROP: BASKETBALL OR RUBBER BALL

[BOUNCE BALL]

(INCREDULOUS VOICE)
Hi. I'm on my way to the bank. They're claiming another check bounced. I tell you, they have some nerve! Calling me up and telling me to **dribble** my checkbook into the bank!

BEEP. . .

PROPS: HOT WATER POT
SPOON
COFFEE CUP

[HOT WATER POT WHISTLING—SHUT OFF—RATTLE A SPOON IN A COFFEE CUP]

Oh, hi! I'm off to work so leave a message and I'll call ya when I get back home.
[SIPPING NOISE?]
Ahhhh! Ya know, nothing gets ya started in the morning like a good hot cup of **beer**!

BEEP. . .

PROP: TV

[TV TUNED TO "WHEEL OF FORTUNE"]

Hi. This is PAT. I can't get to the phone. I'm watching "Wheel of Fortune". Wait! I think I got it—it's a phrase! Yeah! "Leave your name and number at the tone." Yeah! Yeah! That's it, "leave your name and number at the tone!"

BEEP. . .

("SURFER" VOICE)

Hey dude, it's, like, my birthday and here's a list of what you can get me:
1) A new surfboard.
2) Some Coppertone.
3) A knarly beachtowel.
But, like, don't buy me a book, dude—I already have one.

BEEP. . .

PROPS: TIN CUP or CAN
COINS

[DROP QUARTER INTO TIN CUP]
Hi. This is PAT. . . Hang on a second. . .
[DROP COIN INTO CUP]
I have one of those new coin-operated answering machines. Damn thing is costing me a fortune!
[DROP COIN INTO CUP]
Anyway, at the tone please leave. . .
(PINCH NOSE)
"Please deposit another sixty cents if you wish to continue your message!
(NORMAL VOICE)
Forget it Operator! I'm not about to put in another nickel! And if you don't like it, you can. . .

BEEP. . .

PROP: TYPEWRITER

[PECK TYPEWRITER KEYS—ONE FINGER METHOD: "CLICK-CLICK . . . CLICK . . . CLICK-CLICK, ETC.]

Hello. You have reached the new, **experimental** "Answer-Writer". I'm a combination answering machine/typewriter. As you leave your message, I type it out for my owner to read. Get ready for the "beep"—oh, and do me a favor. . . **please speak slowly**!

BEEP. . .

(TWO VOICES—SIMULTANEOUSLY)
Hi. This is **NAME** and **NAME** speaking to you in all new 'Answering Machine Stereo'! We're both taking off right now, however, **one** of us will return your call later—but probably in **mono**.

BEEP. . .

PROPS: TOASTER
OVENTIMER

[PUNCH TOASTER AND START TIMER SIMULTANEOUSLY—"TICK-TICK-TICK, ETC."]

(SPEAK FAST)
Hi. I had my machine fixed recently and the repairman used some old toaster parts. Anyway, hurry up and leave your name and number 'cuz every 30 seconds the tape pops out!

BEEP. . .

(4-YEAR-OLD VOICE)
Hi. Leave a message—and maybe you can help me out? My Mommy read me a nursery rhyme and I don't get it??? Maybe you can tell me—if "Mary had a little lamb. . ." who was the father?

BEEP. . .

**PROPS: WOOD RASP — OR HEAVY-
          GAUGE SANDPAPER
          SCRAP BOARD
          HANDSAW**

[FILE OR SAND A BOARD]

(PETITE VOICE)
Hi. Please leave a message and I'll get back to
you soon. I can't pick up 'cuz I'm in the middle
of doing my nails.
[STOP FILING]
Darn! Look at that hangnail! I better cut it!
[SAW BOARD]
Almost off. . .
There, got it!
[DROP BOARD TO FLOOR]

BEEP. . .

PROP: TV

[CHANGING CHANNELS ON TV—STOP ON COMMERCIAL]

Hi. PAT here. I'm not answering the phone 'cuz I'm watching as many commercials as I can this morning. I'm trying to build up an immunity so tonight when I'm watching my favorite show and they interrupt for a commercial message, it won't make me sick.

BEEP. . .

(PROFESSIONAL VOICE)
Good day. Scientists have recently made a direct connection between the fear of talking to a machine and sexual inadequacy. If you hang up, *I* won't know who you are, but *you'll* know what your problem is.

BEEP. . .

PROP: TAPE RECORDER

[PLAY THE PRE-RECORDED OPENING
BARS OF *THE TONIGHT SHOW* THEME]

Heeeeeeeers. . .

BEEP. . .

[BANG ON MACHINE THROUGHOUT]

Hey, listen—when you leave your message, you
can scream, you can curse, you can abuse this
machine all you want—cuz *my* machine's in the
shop—this is a *loaner*!

BEEP. . .

PROP: GROCERY BAG

[TALK WITH GROCERY BAG OVER HEAD—
RUSTLE BAG]

(MUFFLED VOICE)
Hi. I'm trying to win a bet. Leave a message and
I will call you back as soon as I punch my way
outta here!
(ASIDE)
Hey, this ain't fair—I don't think this bag's wet
enough!

BEEP. . .

PROP: 2 OR MORE BEER OR SOFT DRINK CANS—UNOPENED

(WOMAN SPEAKING IN DEEP MAN'S VOICE)
Hi. This is PAT. I accidentally put on my husband's *manly* deoderant. Anyway, I can't pick
up the phone 'cuz there's an important game
on. Leave a message and I'll get back to ya after
I chug this-here six pack!

[POP OPEN TWO OR MORE CANS]

BEEP. . .

PROP: TOILET

[TAIL-END OF TOILET FLUSH IN BACKGROUND]

Hi. This is PAT. I got a real big problem. Maybe you can help me??? I can't figure out why the *blue water* in my toilet keeps turning green???

BEEP. . .

(FIGHT ANNOUNCER VOICE)
"It's a right to the jaw. . . A left jab to the head. . . It's getting ugly, folks! Whoa, there goes a vicious uppercut. . . I can't believe they're letting this go on! There's blood everywhere!
(NORMAL VOICE)
Oh, hi! This is PAT. We're in the middle of a family discussion—so please leave a message at the *bell*—I mean, the *beep*.

BEEP. . .

PROPS: STEREO
PATRIOTIC MUSIC

[PATRIOTIC MUSIC IN BACKGROUND—e.g. "BATTLE HYMN OF THE REPUBLIC"—AT LOW VOLUME. GRADUALLY TURN UP]

Before you hang up, I just want to remind you that there are certain communist countries where people don't have a choice—because 99% of the people don't have machines—and what machines there are, are controlled by the government. So here's your chance to exercise your God-given freedom of speech! When you hear the beep. . .
(BOSTONIAN ACCENT)
"Ask not what the machine can do for you—but what you can do for the machine!"

BEEP. . .

PROP: OVENTIMER—OR LOUD TICKING CLOCK

[START CLOCK TICKING]

Welcome to "30 Seconds", the answering machine news program, with Mike Wallet and Harry Reasonable. Unfortunately, today we only have time for Andy Fooey:
[TURN OFF CLOCK]
(WHINY VOICE)
"Don't you hate it when you call someone and reach a machine? Why is it people *have* these machines? I hate machines! And did you ever notice that. . .

BEEP. . .

(NASAL VOICE)
Greetings, Earthlings. This is ENAK, speaking to you from the planet Zeblakon. We have taken PAT to our planet for experimentation. We will bring him back at approximately six o' clock Earth time—well, depending on the rush-hour traffic.

BEEP. . .

**PROPS: STEREO
HEAVY METAL MUSIC**

(MOROSE)
Hi. I'm really depressed today so I'm not answering the phone. A friend of mine told me the best way to get rid of depression is to listen to your favorite music. So I'm cueing-up Side One of the group "Suicide Tendencies".

[OPENING BARS OF HEAVY METAL ROCK MUSIC]

BEEP. . .

Hi. This is PAT. Leave a message. I'm outta here.
(NEW VOICE—CONSPIRATORIAL WHISPER)
Hi. This is PAT's machine. Now that he's gone we can talk about him! Boy, you know what I hate? First thing every morning when he speaks into my mike—talk about bad breath! So—what bugs *you* about him?

BEEP. . .

PROPS: STEREO
POLKA MUSIC

[OPTIONAL: PLAY "POLKA MUSIC" IN BACKGROUND]

(SINGING)
"Put your left foot in, put your right foot out, do the Hokie-Pokie and. . ."
(SPEAKING)
Hey! How ya doin? I can't answer the phone right now. We're having a **wild** and **crazy** bachelor party here! Call ya later.
(SINGING)
"Put your left foot in. . ."

BEEP. . .

Hello. Please leave a message. I'm in a meeting with my financial advisor.
("STREET" VOICE IN BACKGROUND)
Yo! PAT! Have I ev-a steered ya wrong? Take my advice, pal! Quicksilver ta Place in the 5th! Safe bet, my man!

BEEP. . .

PROP: DECK OF CARDS OR NEW BOOK

[SHUFFLE CARDS OR RUFFLE PAGES OF NEW BOOK]

Hi. If it's my girlfriend calling, uh, got my new Psych. 101 textbook and I'll be breaking it in all night! Call ya tomorrow, Hon!
(WHISPER)
If it's the guys—got a brand new deck and I'll have it broke in by game time tonight!

BEEP. . .

(AUSTRIAN ACCENT)
This is a powerful Arnold Schartzinegger Answering Machine. If you don't leave a message your phone service will be *terminated*!

BEEP. . .

PROPS: STEREO
CLASSICAL MUSIC

[OPERATIC OR CLASSICAL MUSIC IN BACKGROUND]

(STUFFY VOICE)
Please be so kind as to leave a message, as it's off to the opera for me. I mean, what a simply charming way to top off a ravishing repast of **pate' foie gras** appetizer, **champions le' beuf** entree and **le' flambe** dessert with vintage **Dom Perignon**!

[LET OUT LOUD BELCH]

BEEP. . .

(VOICE FAR IN BACKGROUND)
Hello! You'll have to leave a message. I'm waiting for the Fire Department to get me out of this tree! Mark my words, this is the last time I go out on a limb for my wife!

BEEP. . .

**PROPS: HAND SAW
        SCRAP BOARD**

Hi. I won't be able to answer the phone. Some-
one just arrived at my house and it's another
chance for me to try this magic trick I keep
messin' up. . .
[SAW BOARD]
(SCREAM!)
Hey, come back!
(TO CALLER)
Say, uh, listen—you doin' anything this
afternoon?

BEEP. . .

**PROPS: HANDFUL OF COINS
        TIN CUP OR POT**

(EXCITED)
PAT here! Don't hang up 'cuz I got this great new
answering machine in Las Vegas—and
sometimes when you leave a message, it pays
off! I'm gonna try it!
This is PAT. My phone number is 123-4567. . .
[POUR COINS INTO TIN CUP]
Wow! Jackpot! Hey, are *you* feelin' lucky today?

BEEP. . .

PROP: SINK FULL OF WATER

[RUN WATER IN SINK—SHUT OFF—MAKE *SPLASH* NOISE, SIMULATING SOMETHING HEAVY DROPPED INTO THE WATER]

(SPEAK WITH WATER IN YOUR MOUTH, MAKING A GARGLING SOUND)
Hi. This is PAT. Please help me out and leave a message. I'm testing out my new water-proof answering machine!

BEEP. . .

PROPS: STEREO
BEATLE RECORD

[PLAY A *BEATLE* RECORD IN THE BACK-GROUND]

Hi. I probably won't hear the phone when it rings—I'm in the other room listening to a tape of an old Rock group a friend of mine gave me. He said they broke up a long time ago.
(TO SELF)
Boy, you call *that* music!!??
No wonder they never made it!

BEEP. . .

(MECHANICAL VOICE)

Hello. I'm a computer. . . What are you doing?
. . . Want to go out with me? . . . Before you say
no—just answer me this. . . Have you ever ***tried***
computer dating?

BEEP. . .

PROP: EXCERCISE BIKE

[PEDAL EXERCISE BIKE—HEAVY BREATH-
ING IN BACKGROUND]

Hi. I'm home but I'll have to get back to ya. My
doctor says he wants 10 miles on my exercise
bike odometer every day. . .
(TO PERSON IN BACKGROUND)
Faster, Consuella, faster!

BEEP. . .

TITLES BY CCC PUBLICATIONS
— NEW BOOKS —

NEVER A DULL CARD
WORK SUCKS!
THE UGLY TRUTH ABOUT MEN
IT'S BETTER TO BE OVER THE HILL — THAN UNDER IT
THE PEOPLE WATCHER'S FIELD GUIDE
THE GUILT BAG (Accessory Item)
HOW TO **REALLY** PARTY!!!
THE ABSOLUTE **LAST CHANCE** DIET BOOK
HUSBANDS FROM HELL
HORMONES FROM HELL (The Ultimate *Women's* Humor Book!)
FOR **MEN** ONLY (How To Survive Marriage)
THE Unofficial WOMEN'S DIVORCE GUIDE
HOW TO TALK YOUR WAY OUT OF A TRAFFIC TICKET
WHAT DO WE DO NOW?? (The Complete Guide For All New Parents

— COMING SOON —

THE BOTTOM HALF
LIFE'S MOST EMBARRASING MOMENTS
HOW TO ENTERTAIN PEOPLE YOU HATE
THE KNOW-IT-ALL HANDBOOK

— BEST SELLERS —

NO HANG-UPS (Funny Answering Machine Messages)
NO HANG-UPS II
NO HANG-UPS III
GETTING EVEN WITH THE ANSWERING MACHINE
THE SUPERIOR PERSON'S GUIDE TO EVERYDAY IRRITATIONS
YOUR GUIDE TO CORPORATE SURVIVAL
GIFTING RIGHT (How To Give A Great Gift On Any Budget!)
HOW TO GET EVEN WITH YOUR EXes
HOW TO SUCCEED IN SINGLES BARS
TOTALLY OUTRAGEOUS BUMPER-SNICKERS
THE "MAGIC BOOKMARK" BOOK COVER (Accessory Item)

— CASSETTES —

NO HANG-UPS TAPES (Funny, Pre-recorded Answering Machine
    Messages With Hilarious *Sound Effects*) — In Male or Female Voices
Vol I: GENERAL MESSAGES
Vol II: BUSINESS MESSAGES
Vol III: 'R' RATED MESSAGES
Vol IV: SOUND EFFECTS ONLY
Vol V: CELEBRI-TEASE (Celebrity Impersonations)